ANIMALS THAT GLOW
DEEP-SEA FISH

by Joyce Markovics

Cherry Lake Press
Ann Arbor, Michigan

Cherry Lake Press

Published in the United States of America by Cherry Lake Publishing Group
Ann Arbor, Michigan
www.cherrylakepublishing.com

Reading Adviser: Beth Walker Gambro, MS Ed., Reading Consultant, Yorkville, IL
Content Adviser: Steven Haddock, Senior Scientist and Marine Biologist
Book Designer: Ed Morgan

Photo Credits: © SamRobertshaw/Shutterstock, cover; © Neil Bromhall/Shutterstock, title page; © freepik. com, copyright page and TOC; © Pally/Alamy Stock Photo, 4-5; CSIRO, Wikimedia Commons, 5; © Joern_k/Shutterstock, 6; Chip Clark/Smithsonian Institution, Wikimedia Commons, 7; © Andy Murch/naturepl.com, 8; © Doug Perrine/naturepl.com, 9; © Pally/Alamy Stock Photo, 10-11; © EtiAmmos/Shutterstock, 12; © Kelvin Aitken/VWPics/Alamy Stock Photo, 13; © Doug Perrine/Alamy Stock Photo, 14; © Pally/Alamy Stock Photo, 15; © Neil Bromhall/Alamy Stock Photo, 16; © Solvin Zankl/Alamy Stock Photo, 17; © Pally/Alamy Stock Photo, 18-19; © Steven Haddock biolum.eemb.ucsb.edu, 20; © Steven Haddock biolum.eemb.ucsb.edu, 21; © Kletr/Shutterstock, 22 top left; © Pally/Alamy Stock Photo, 22 top right; © Mirt Alexander/Shutterstock, 22 bottom.

Copyright © 2022 by Cherry Lake Publishing Group
All rights reserved. No part of this book may be reproduced or utilized in any form or by any means without written permission from the publisher.

Cherry Lake Press is an imprint of Cherry Lake Publishing Group.

Library of Congress Cataloging-in-Publication Data

Names: Markovics, Joyce L., author.
Title: Deep-sea fish / by Joyce Markovics.
Description: Ann Arbor, Michigan : Cherry Lake Publishing, [2022] | Series: Lights on! animals that glow | Includes bibliographical references and index. | Audience: Grades 4-6
Identifiers: LCCN 2021035054 (print) | LCCN 2021035055 (ebook) | ISBN 9781534199590 (hardcover) | ISBN 9781668900734 (paperback) | ISBN 9781668906491 (ebook) | ISBN 9781668902172 (pdf)
Subjects: LCSH: Deep-sea fishes—Juvenile literature.
Classification: LCC QL620 .M36 2022 (print) | LCC QL620 (ebook) | DDC 597.177—dc23
LC record available at https://lccn.loc.gov/2021035054
LC ebook record available at https://lccn.loc.gov/2021035055

CONTENTS

Glowing Sharks................4
Predators of the Deep...........10
Alien Anglerfish................14
Fish Flashlights................18

Creature Feature 22
Glossary......................23
Find Out More..................24
Index 24
About the Author................24

GLOWING SHARKS

In 2020, scientists made a shocking discovery off the coast of New Zealand. "I nearly cried when I saw it . . . it was so exciting," lead scientist Jérôme Mallefet said. Before his eyes was a glowing kitefin shark. Its skin was giving off a soft blue-green light.

The big-eyed kitefin shark

Jérôme and his team found the guitar-size shark in the ocean's twilight zone. Very little light can reach this area. It extends 3,000 feet (914 meters) underwater. This dark place is home to many animals that have an amazing adaptation—they produce light!

Jérôme Mallefet and his team worked on a research boat like this one when they made the kitefin shark discovery.

Bioluminescence (buy-oh-loo-muh-NES-uhns) means "living light." It's a chemical process that allows animals to make their own light.

The kitefin shark is one of several sharks that are bioluminescent. More than 10 percent of sharks glow, says Jérôme. However, the kitefin is the largest of them. It's also the biggest **vertebrate** ever discovered that produces its own light!

Sharks have long bodies and often hunt other animals for food.

Deep-sea lantern sharks also make light. The smallest is the dwarf lantern shark. It can fit in a person's hand. Lantern sharks have tiny spots called photophores along their bellies and sides. Photophores make light as a result of a chemical reaction.

A dwarf lantern shark feeds on small fish, squid, and shrimp.

In most animals, bioluminescence occurs when a substance called luciferin (loo-SIH-fuh-rin) mixes with oxygen. This chemical reaction creates light. However, scientists aren't exactly sure of all the chemicals sharks use to glow.

Velvet belly lantern sharks

Lantern sharks use their light to hide from **predators**. Like kitefin sharks, they live in deep, dark ocean waters where there is little light. But if a hungry predator spots the shark from below, there's just enough light to make it visible and **vulnerable** to attack.

One kind of lantern shark has two spines on its back. These glow like light **sabers**, warning enemies to keep away!

The glow from lantern sharks' bellies matches the faint sunlight coming from above. So the fish appear invisible to predators! In other words, the sharks "don't show their shadows," says Jérôme. This keeps them safe from attackers. The light could also help them attract smaller animals to eat.

Some scientists think lantern sharks use their glowing bellies to communicate with other lantern sharks and find mates.

PREDATORS OF THE DEEP

Sharks aren't the only glow-in-the-dark predators in the sea. Viperfish thrive in the twilight zone. And they look like monsters, with huge needles for teeth. In fact, their teeth are so long and pointy that they can't close their mouths! But the animals themselves aren't that scary. They rarely grow more than 1 foot (30 centimeters) long.

Like other viperfish, Sloan's viperfish has photophores along its body. They sparkle like tiny jewels. This viperfish uses its glow to hide from predators and to attract prey. When the prey comes near, the viperfish **impales** its victim on its sharp teeth. *Snap*! The viperfish can then **unhinge** its jaws to swallow an animal 60 percent bigger than itself.

The viperfish's favorite food is lanternfish, another bioluminescent fish. However, it will pretty much eat whatever it can catch. Scientists have found fish eggs and **crustaceans** inside viperfish stomachs.

Viperfish have light-up parts all over, even on their fins and around their eyes!

Sleek and silver, another viperfish has fangs that shoot past its eyeballs. The foot-long (30-cm) Pacific viperfish swims to shallower waters to feed at night. After it snaps up a shrimp or other prey, its teeth form a cage around the prey. Then it swallows its meal whole.

A Pacific blackdragon

The Pacific blackdragon is another toothy deep-sea predator. It has a long, glowing lure on its chin to attract prey.

ALIEN ANGLERFISH

One of the bright stars of the deep is the anglerfish. Its famous feature is a glowing lure on its snout. Unlike sharks and viperfish, the anglerfish gets its light from millions of bioluminescent bacteria in its lure! When the tiny life forms in the lure glow, they attract prey that the anglerfish gulps down.

An anglerfish can eat prey twice its size!

The relationship between the anglerfish and bacteria is called **symbiosis**. In exchange for light, the anglerfish gives the bacteria a safe place to live. The bacteria also get **nutrients** they need from the fish.

More than 200 kinds of anglerfish exist. Each one appears to be paired with a unique group of bacteria.

In some anglerfish **species**, the males are **parasites**! They are much smaller than females and have tadpole-like bodies. When a male anglerfish finds a female, he bites her. Then he releases a chemical that **dissolves** her skin!

The lumps on top of the female are two smaller males attached to her body.

A tiny male anglerfish attached to a female.

The male and female's bodies **fuse** together. The male feeds on her blood. In return, he gives her the special cells she needs to have young. Sometimes, multiple males clamp onto a single female. They remain with her as long as she is alive.

FISH FLASHLIGHTS

One bioluminescent fish shines extra bright. The flashlight fish lives in shallower waters than most glowing fish. It has a bean-shaped organ under each eye that produces light. Like anglerfish, the organ is filled with bioluminescent bacteria.

Experts think flashlight fish may also use their light to find prey and communicate with other members of their species.

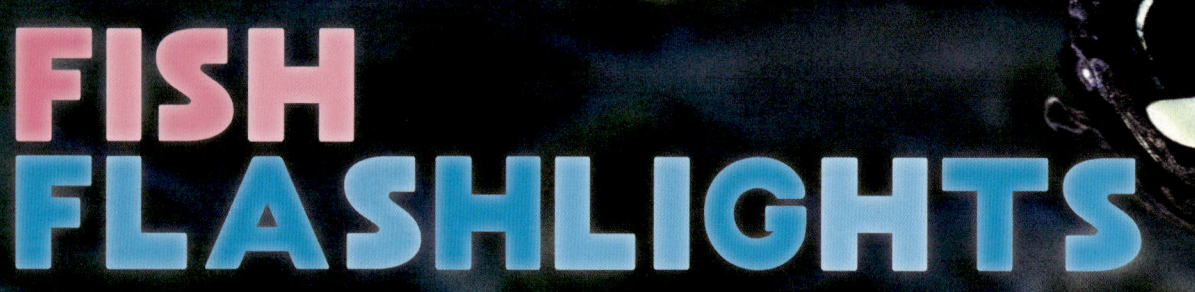

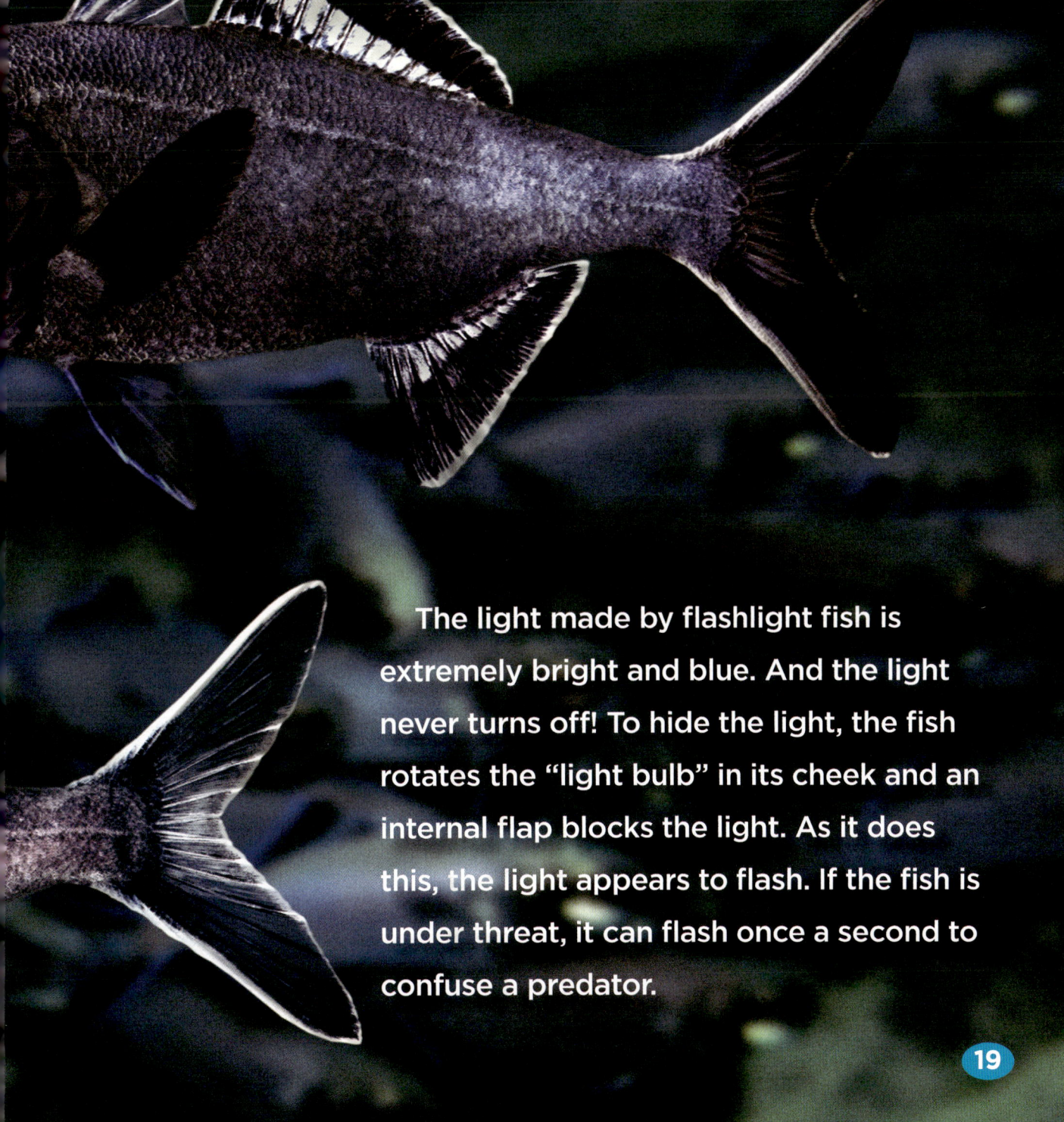

The light made by flashlight fish is extremely bright and blue. And the light never turns off! To hide the light, the fish rotates the "light bulb" in its cheek and an internal flap blocks the light. As it does this, the light appears to flash. If the fish is under threat, it can flash once a second to confuse a predator.

Lanternfish also light their own way. Their silver bodies are covered with glowing dots. They live in deep waters. Although at night, lanternfish swim closer to the ocean's surface to feed. However, they need to keep their big eyes open. Squid, sharks, whales, and seabirds all hunt them.

A lanternfish's big eyes help it see in dark ocean waters.

Lanternfish are very common, but they're still at risk like all deep-sea bioluminescent animals. Climate change and pollution are the biggest threats. "Down there, there are glowing critters of different sizes . . . that we still know nothing about," says Jérôme Mallefet. "It's time to study this ecosystem before we destroy it."

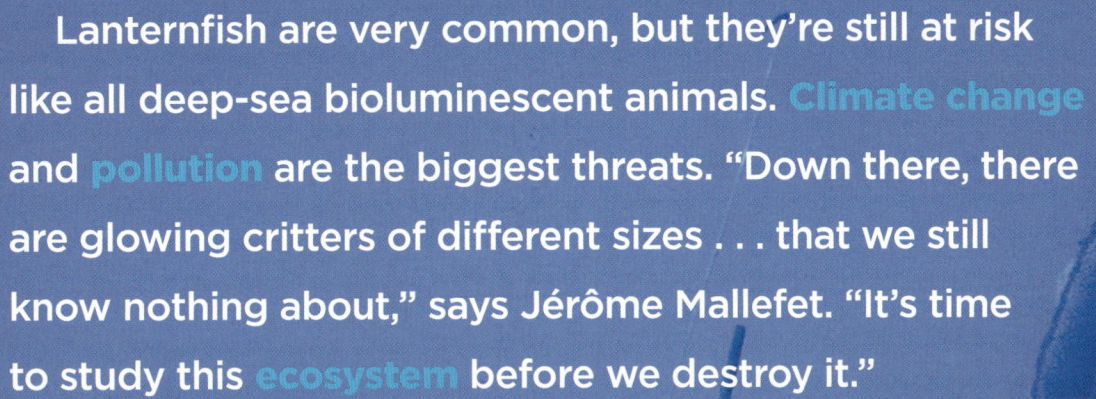

Scientists believe lanternfish use their lights to stay safe and attract mates.

CREATURE FEATURE

HATCHETFISH

- Hatchetfish live in the deep sea. They have huge bulging eyes on top of their heads that help them spot prey in the dark.

- A hatchetfish is flat like a coin and can fit in your hand. It earned its name because its body is thin like the blade of a hatchet

- Hatchetfish produce blue light along their bellies and sides. They have lenses and reflectors to direct the light. This helps hatchetfish appear invisible from below, keeping them safe from hungry enemies.

GLOSSARY

climate change (KLYE-muht CHAYNJ) a long-term change in Earth's climate, especially an increase in the temperature of the air and oceans

crustaceans (kruh-STAY-shuhnz) groups of animals, such as lobsters or shrimp, that have hard outer skeletons and jointed legs

dissolves (di-ZOLVES) mixing with another substance and turning into a liquid

ecosystem (EE-koh-sih-stuhm) a community of animals and plants that depend on one another to live

fuse (FYOOZ) to blend or become attached

impales (im-PAYLZ) stabs with something sharp

lure (LOOR) something that attracts an animal

nutrients (NOO-tree-uhnts) substances needed by living things to grow and stay healthy

parasites (PA-ruh-sites) living things that live and feed on another living thing

pollution (puh-LOO-shuhn) harmful materials that damage the air or water

predators (PREH-duh-turz) animals that hunt other animals for food

reflectors (ri-FLEK-tuhrz) objects that return light

sabers (SAY-buhrs) swords

species (SPEE-sheez) types of animals or plants

symbiosis (sim-bee-OH-suhs) when two different organisms live together in a relationship that benefits both

unhinge (uhn-HINJ) to open wide

vertebrate (VUR-tuh-bruht) an animal with a backbone

vulnerable (VUHL-nuh-ruh-buhl) able to be easily hurt

FIND OUT MORE

Books

Davidson, Rose. *Glowing Animals*. Washington, DC: National Geographic, 2019.

Martin, Lee Anne. *Amazing World: Sea Creatures*. Bellevue, WA: Quarto Publishing, 2017.

Regan, Lisa. *Way to Glow! Amazing Creatures That Light Up in the Dark*. New York, NY: Scholastic, 2016.

Websites

Monterey Bay Aquarium: Hatchetfish
https://www.montereybayaquarium.org/animals/animals-a-to-z/hatchetfish

NOVA Science Now: Glowing in the Dark
https://www.pbs.org/wgbh/nova/sciencenow/0305/04-glow-nf.html

Woods Hole Oceanographic Institution: Ocean Twilight Zone
https://twilightzone.whoi.edu/explore-the-otz/creature-features/sloanes-viperfish/

INDEX

anglerfish, 14–18
bacteria, 14–15, 18
bioluminescence, 5–7, 14, 18, 21
climate change, 21
crustaceans, 12
ecosystem, 21
flashlight fish, 18–19
hatchetfish, 22
lanternfish, 12, 20–21
luciferin, 7
Mallefet, Jérôme, 4–5, 21

Pacific blackdragon, 13
photophores, 7, 11
pollution, 21
predators, 11, 13, 19
prey, 11, 13–14, 19, 22
sharks
   lantern, 6–9
   kitefin, 4–6, 8
teeth, 10–11, 13
twilight zone, 5, 10
viperfish, 10–14

ABOUT THE AUTHOR

Joyce Markovics loves writing about animals that seem too bizarre to be real. The deep sea is a treasure trove of life forms, many of which have yet to be studied. She hopes this book inspires young readers to take action to protect our oceans and the amazing animals in them. She dedicates this book to Michael Paul, a curious explorer.